Panzerkampfwagen 38(t)
Variants-2.0 cm; 3.7 cm; 7.5 cm; 7.62 cm; 15cm

David V. Nielsen

ZIMMERIT
PRESS

1999

❖ ❖ ❖

The author has attempted to verify technical data from a minimum of two sources. It has been discovered that numerous inconsistencies between the sources exist and the numbers chosen in the event that verification was unavailable were those believed to be of greatest reliability.

❖ ❖ ❖

The author gratefully acknowledges the assistance of CKD and the Imperial War Museum and the kind permission granted by those institutions to reprint photographs from their archives.

❖ ❖ ❖

The publisher is looking for related topics and titles and welcomes inquiry from authors.

❖ ❖ ❖

The publisher is interested in any photographs of armor, vehicles, soldiers, weapons, etc.
of any army or nation and is looking for wartime pictures.
All photographs provided will be copied and the original returned.
Full credit will be given to the donor for any photograph which may be used.
These items can be sent to the attention of the publisher at
Zimmerit Press, P.O. Box 1192, Boise, ID 83701.

Table of Contents

I Introduction

The Panzerkampfwagen 38(t) chassis, due to a rugged reliable design, was used as the basis for numerous variants some of which achieved a fame which exceeded that of the original production vehicle. The 38(t) chassis, easily recognizable due to the four large road wheels became a platform for anti-aircraft guns, anti-tank guns, self propelled artillery and munitions carriers. Additionally, a light reconnaissance tank variation produced in early 1942 represented one of the more modern designs of its day but was never placed into production.

The variations presented in this book are organized by weapon size and illustrate the development of the varied roles to which this vehicle was pressed into service. Some variations of course were built upon their predecessors and incorporated modifications which improved upon those designs already in service. Several variations although produced in large numbers and the subject of extensive field service were expedient designs which relied upon the availability of existing weapon stores rather than a deliberate attempt to introduce an original dedicated purpose vehicle.

The first modification made to the 38(t) was the 7.62cm PaK 36(r)(Sd.Kfz 139) which mounted captured Russian 7.62cm guns on a chassis from which the turret had been removed. At this time the 38(t) vehicle had been determined to be obsolete and the conversion provided a quick means of producing self propelled anti-tank vehicles. While crew protection was limited these vehicles saw extensive use in North Africa and the Eastern Front in Russia.

This was followed by the 7.5 cm PaK 40/3 auf PzKpfw 38(t) Ausf H (Sd.Kfz 138) and the 7.5 cm PaK 40/3 Ausf M (Sd.Kfz 138). The model H is designated such for the term *heckmotor,* which indicates a rear-mounted engine. The designation M is for *mitte* as the engine location was moved to the middle of the vehicle. Changes were made to both the gun and placement of the fighting compartments. Modification of the gun shield as well as changing the location of the engine improved this design and represented the apex of these vehicles. Combat service saw them used on both fronts and their large production numbers attest to their performance and function. These three variants alone would secure a place in history for these vehicle types yet an additional development, namely that of the self-propelled artillery variant, also played a significant role in the German armored forces.

Two 15-cm variations, the 33 Ausf H (Sd.Kfz 138/1) and 33/1 Ausf M (Sd.Kfz 138/1) were used on both the Eastern and Western fronts. The first is distinguished by the large enclosed fighting compartment which encompassed almost the entire surface area of the chassis. The second placed the gun towards the rear of the vehicle with a smaller fighting compartment. The limited ammunition storage of these vehicles necessitated the production of a munitions carrier which was to serve alongside the artillery.

The Flakpanzer variant of the 38(t) incorporated an armored superstructure which used fold-down plates. The fighting compartment was located at the rear of the vehicle. The single 2cm gun, however, proved to be of limited performance.

An additional intriguing variation of the 38(t) was the 2 cm KwK 38, a reconnaissance vehicle that mounted the 2 cm Hangelafette 38 turret which was also used on the Sd.Kfz 222 armored car.

The scope of this publication is of course limited and is not by any means intended as a full history of the many intriguing variants of what proved to be an extremely capable and adaptable original vehicle design. The author would refer those individuals who wish to investigate these vehicles at length to read the outstanding works by Walter J. Spielberger, Charles Kliment and Vladimir Francev for additional information.

Panzerkampfwagen 38(t)

II Flakpanzer 38(t) auf Selbstfahrlafette 38(t) Ausf M (Sd.Kfz 140)

Vehicle Type: Flakpanzer 38(t) auf Selbstfahrflafette 38(t) Ausf M (Sd.KFz 140)
Manufacturer: BMM
Production Number: 140
Armament: 2 cm Flak 38 L/112.5
Crew: 4
Weight: 9.8 tons
Length: 4.86 meters
Width: 2.14 meters
Height: 2.25 meters
Armor: 10 - 15 mm
Engine Type: Praga EPA - 2, 6 cylinder
Speed: 42km/hr.
Range: 210 km

The 2 cm Flak 38 was a self propelled anti-aircraft gun mounted in a rear fighting compartment. The superstructure on the fighting compartment was hinged to fold down and provide full elevation and traverse of the weapon. Interrupter bars were fixed on the vehicle directly in front of the fighting compartment. The vehicle carried 1,040 rounds of ammunition and utilized a crew of 4-5. Production of the vehicle started in November of 1943 and continued until February of 1944.

Right side view. Note vehicle exhaust location. Hinged armor plate in upright position. Gun in travel position. Factory photograph indicates position of external vehicle tool storage.

Overhead rear view. Driver's hatch padding visible. Crew seat in down position. Good illustration of spent shell casing container. Tetra fire extinguisher located on rear left side of vehicle. Tow cables shown individually wrapped rather than using overlapping format.

Overhead rear view. Spare barrel container in 12:00 position immediately in front of gun. Ammo and storage boxes in place. Heater vent located on front right side of compartment. Note elevation crank assembly and sight bracket assembly.

Forward view of fighting compartment. Sight mounting bracket clearly visible as well as clamps in place between armor plates. Canvas cover on barrel. View of interrupter bars located immediately below barrel. Also note radio rain cover. Spare barrel box indicates 2 cm flak 38 Rohr Waffe Nr.

Overhead view 12:00 position. Spent shell case receptacle visible. Ammo boxes in place around pedestal. Gunner seat has been removed. Hand traverse wheel at 6:00 position. Note lifting handles on diamond floor plating and the attachment point for gun shields. Safety chain attached to eye bolts for gun shields.

Overhead view of fighting compartments. Gun placed in 3:00 position. Trunion mount in forward position. Note interior gas mask container bracket. Back sight is missing.

Two views of same vehicle with different gun elevations. Below: Left hand side three quarter view. Armor plates in upright position. Gun appears to be in fully raised position. Driver's vision port raised in open position. Vehicle is not fully fitted as external storage items and flak sights are missing. Right: Right side three quarter view. Gun at zero degrees or ground attack position.

Below: Overhead right side three quarter view. Good view of interrupter bar. Gun cleaning rod position clearly observed as well as tool storage. Notek light in place. View of spent shell casing receptacle attached to gun.

Left side view. Armor plates in lowered poition. Wooden jack block stowed above first road wheel with toolbox behind. Jack mounted over toolbox. Sprocket wheel of semi perforated design.

Above: Left side three quarter front view. All armored plates in down position. Gun at max elevation. Note driver's padded split hatch opened.

Left: Rear view. Stowage of tow cable in circular non overlapped configuration. Fire extinguisher in place on left hand side.

III Aufklarer auf Fahrgestell Panzerkampfwagen 38(t) mit 2cm KwK 38 (Sd.Kfz 140/1)

Vehicle Type: Aufklarer auf Fahrgestell Panzerkampfwagen 38(t) mit 2cm KwK 38 (Sd.Kfz 140/1)
Manufacturer: BMM
Production Number: 50
Armament: 2cm KwK38 L/55
7.92mm MG42
Crew: 4
Weight: 9.5 tons
Length: 4.51 meters
Width: 2.14 meters
Height: 2.17 meters
Armor: 8 - 50 mm
Engine Type: Praga EPA
Speed: 58km/hr.
Range: 250 km

The Aufklarungspanzer was a reconnaissance vehicle which was assembled on units returned for repair. The conversion entailed mounting a 2 cm Hangelafette 38 turret on the standard 38(t) chassis. The first conversions undertaken in February and March of 1944 started with a modest 4 vehicles. The vehicle was fitted with anti-grenade screens. Production ended in March of 1944. A total of 50 units were made with two additional units being tested with the 7.5 cm KwK 38 L/24 gun.

Right hand side three quarter front view. Anti-grenade screening in open position on turret. Vehicle missing vision port cover for driver and transmission hatch partially open. Antenna mount located on front immediately below tool stowage box.

Head on view of same vehicle. Note absence of bullet splash guard. Accumulation of mud and debris indicates either testing or field service.

Rear right three quarter view. Workshop photograph. Rear turret stowage boxes reinforced by metal strapping. Note also placement of fire extinguisher. Antenna fitted. Idler wheel has pear or teardrop shaped hole configuration.

Front left three quarter view. Note smoke dischargers. New style bow machine gunner/driver plate which eliminated the bow machine gun plate aperture.

Illustration of likely chassis test with turrets not yet in place on vehicles. Note wind screen plates in place for testing. Vehicles equipped with two Notek lights and test driving headlight.

IV Panzerkampfwagen 38(t) neuer Art

Vehicle Type: Panzerkampfwagen 38(t) neuer Art
Manufacturer: BMM
Production Number: 15
Armament: 3.7cm KwK 38(t)
7.92mm MG37
Crew: 4
Weight: 14.8 tons
Length: 4.62 meters
Width: 2.5 meters
Height: 2.4 meters
Armor: 10 - 50 mm
Engine Type: Praga V-8
Speed: 64km/hr.
Range: 200 km

The TNHnA (neue Ausfuhrung) also known as the Neuer Art was a high speed reconnaissance tank developed as an entry into the Wehrmacht's competition to find a new reconnaissance vehicle. Only 15 of the vehicles are believed to have been produced in the late months of 1941 and early months of 1942. The vehicle incorporated a new welded armor construction for both the hull and turret. The size of the hull was larger than the 38(t) as it contained a new engine, the Praga V-8. The turret design also differed from the 38(t) and was comprised of sloped sides and contained a commander observation cupola. The 38(t) chassis was also modified slightly to incorporate larger diameter running wheels. The vehicle mounted a 3.7 cm KwK gun. The vehicle also incorporated a double muffler design.

Right side view. Notek light in place. Note reinforced sand shield over track with white stripe on shield for infantry visibility. The vehicle also incorporates a spot light on turret. Vision port block of driver in upright position. Stowage boxes, jack stand and crowbar in place.

Left side view. Left side engine compartment hatch and muffler are clearly illustrated.

Picture of prototype test drive to Rad Host Mountains. Unknown date.

Rear overhead view. Rear vision blocks on turret. Note turret does not have bullet splash. Full tool storage, including jack, pick, axe, and tow cable. Traffic distancing device with branch deflector also installed.

Rear overhead view. Turret in 3:00 position to allow engine hatch opening. Note movable shield adjacent to air intake.

Front view. Illustration of turret light as well as siren and Notek light. Transmission access plate clearly visible.

Rear view. Tool storage as well as tow cable clearly visible. Cover in place over PTO. Rear reflectors incorporated.

Interior photographs which are likely taken on production line as assembly of engine components is as of this time incomplete. Note left hand drive position of operating controls and welded internal construction. Below looking aft. View of radiator and fuel tanks. Horizontal line at top of photograph is outline of turret race.

V

Panzerjager 38(t) mit 7.5cm PaK 40/3 Ausf H (Sd.Kfz 138)

Vehicle Type: Panzerjager 38(t) mit 7.5cm PaK 40/3 Ausf H (Sd.Kfz 138)
Manufacturer: BMM
Production Number: 611
Armament: 7.5cm PaK 40/3 L/46
7.92mm MG37(t)
Crew: 4
Weight: 10.8 tons
Length: 5.77 meters
Width: 2.16 meters
Height: 2.51 meters
Armor: 8 - 50mm
Engine Type: Praga AC
Speed: 47km/hr.
Range: 240 km

The 7.5 cm PaK 40/3 Ausf H was also known as the Marder III. This vehicle was a modification similar to that of the 7.62 cm PaK 36(r). The German PaK 40gun which was available starting in May of 1942 was incorporated. Similar to the PaK 36(r), a superstructure was mounted over the center of the 38(t) chassis. The superstructure was larger than that used in the PaK 36(r). Production on the vehicle started in November of 1942. Two hundred and seventy-five were manufactured and another 336 vehicles were constructed as conversions.

Right hand side view. Support poles in place for canvas cover. Good view of stowage boxes, jack block, and jack. Unique track storage located on fender. Travel lock for gun in down position. Note turned up perforated plate used for retaining spent cartridge cases.

Overhead rear view showing fighting compartment. View of radio mount, storage box and the engine access plate located below gun.

Rear view. Note two periscopes and presence of auxiliary pop-up periscopes. View of traverse and elevation hand wheels for gunner. Late war cylindrical travel distancing device located above left fender. *(Photo Courtesy Imperial War Museum)*

Rear view of vehicle on train transport - Czechoslovakian Army, 1945. Vehicle behind is a 38(t).

Front right hand rear quarter view. Rain cover in position over fighting compartment and gun. Muzzle brake cover employed.

North Africa - unknown date. Vehicle captured in Tunis.

View of vehicle loading. Note early style night travel distance indicator. Compartment cover partially installed during movement. Breech cover and muzzle brake cover used.

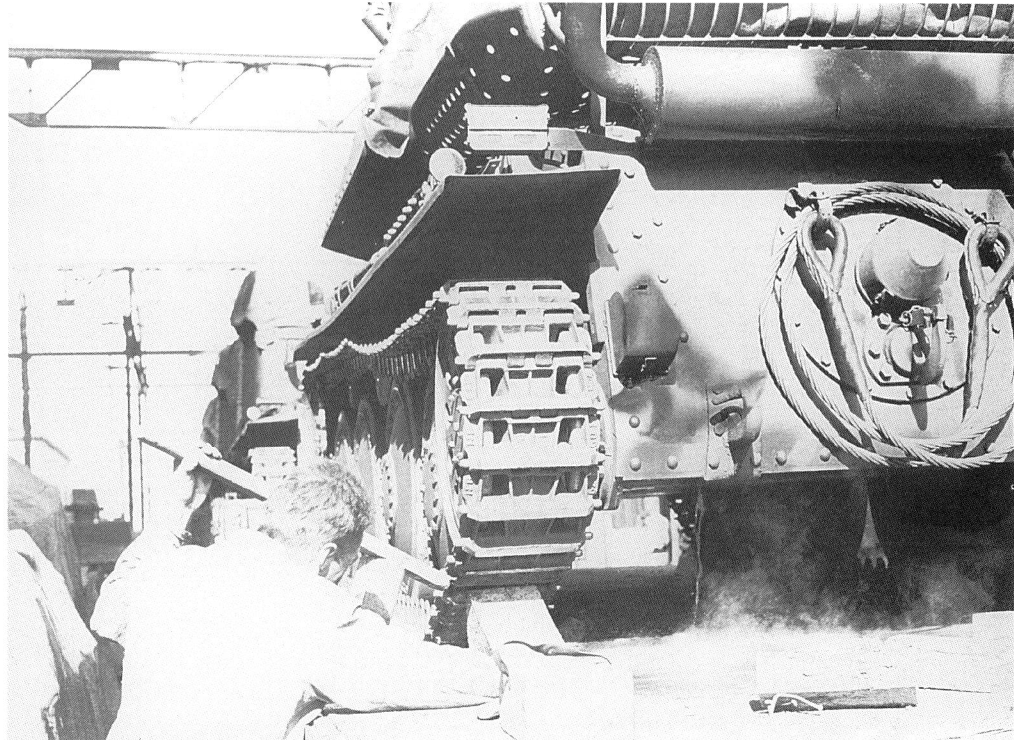

Rear view of loading on train transport. Track being blocked for travel. Square style night travel distance indicator in place.

VI Panzerjager 38(t) mit 7.5cm PaK 40/3 Ausf M (Sd.Kfz 138)

Vehicle Type: Panzerjager 38(t) mit 7.5cm PaK 40/3 Ausf M (Sd.Kfz 138)
Manufacturer: BMM
Production Number: 975
Armament: 7.5cm PaK 40/3 L/46
7.92mm MG34
Crew: 4
Weight: 10.5 tons
Length: 5 meters
Width: 2.15 meters
Height: 2.35 meters
Armor: 10 - 50mm
Engine Type: Praga AC
Speed: 42km/hr.
Range: 200 km

This vehicle was the last anti-tank gun model modification of the series comprising the 7.62 cm PaK 36(r), and the 7.5 cm PaK 40/3 Ausf H. This model incorporated several important changes which mark a significant improvement in vehicle design. The configuration of the fighting compartment was improved as well as having the silhouette of the vehicle lowered. The engine was moved to the front of the vehicle and the fighting compartment as a result was increased in size which in turn provided easier access for early crew members. The driver, who was on the right side, was provided an armored hood, which in early model vehicles, utilized a round cast cover with later model vehicles incorporating a welded design. The gun was moved rearward, and the superstructure now had four sides and extended to the rear of the vehicle. Nine hundred seventy-five units were produced from April, 1943 until May of 1944.

Left front three quarter view. Notek light in place. Clear view of cast driver's hood. Front visor open.

Front view. Note hardware bracket attachment for spare track. Gunners sight aperture plainly visible.

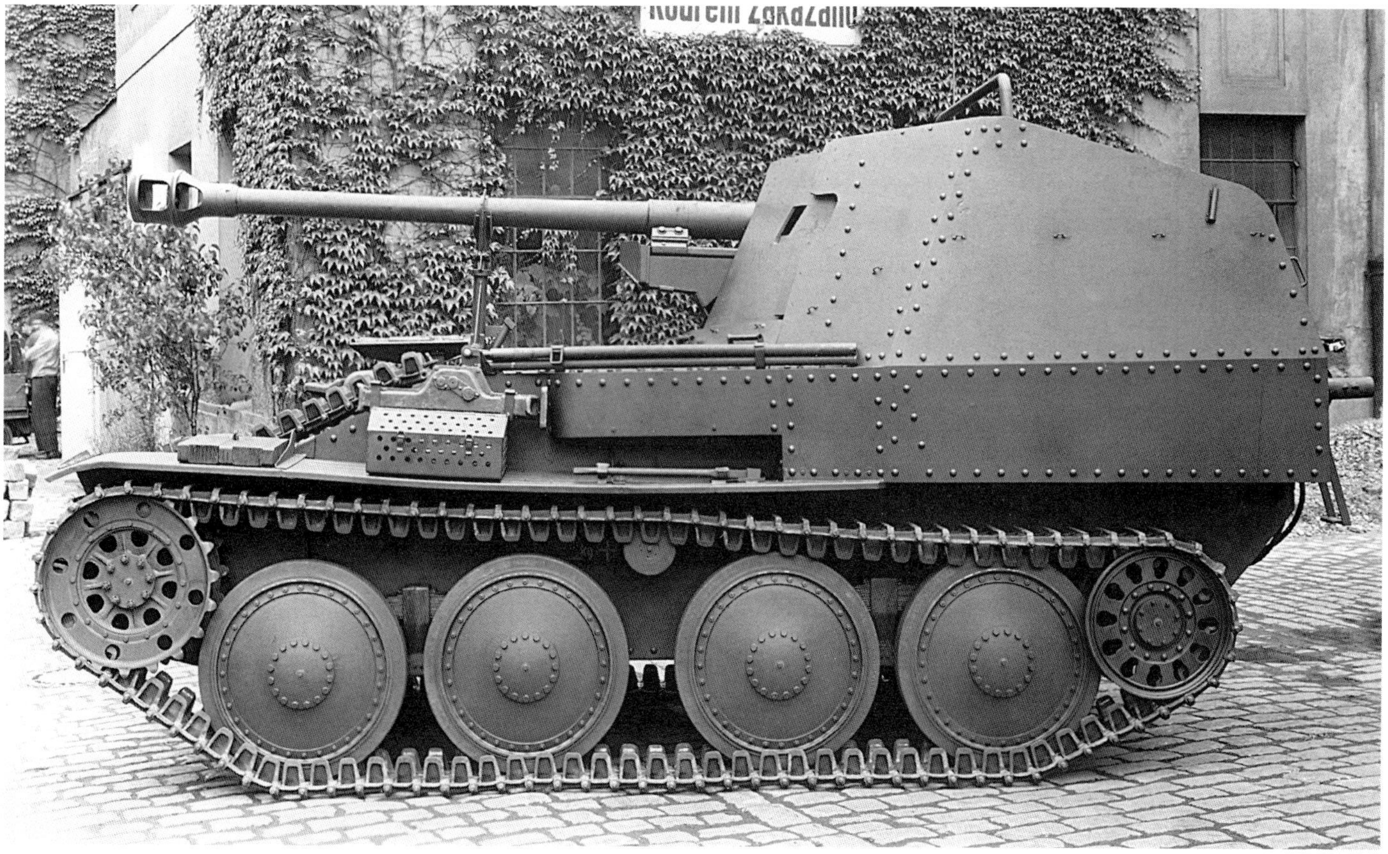

Left side view. Note gun cleaning rods and jack. View of rear step for crew access.

Right side view. Driver's side vision port on right side. Right side view also illustrates gun cleaning equipment and antenna placement.

Front three quarter view.

Right rear three quarter view. Rear armor in up position. Note perforated muffler cover. Crossbar for canvas cover located over fighting compartment. Cylindrical style night position guide in place.

Right rear overhead view. Rear armor plate lowered and clear illustration of slide locks to secure plate. Note rear travel lock for gun. Canvas covers in place over ammo tubes and radio. Circular driver's hatch and square engine compartment hatches are visible.

Overhead rear view. Note semi-circular reinforcing plate located above gun. Spare gunner's sight in place. Gunner's traveling seat secured by clamp mounts. Ammo tubes are located on both sides of compartments. Rear travel lock in secured position.

Two separate pictures from Italy showing knocked out M's. Above from 3rd of June, 1944. Fifth Army - Lanuvio area, Italy. Vehicles knocked out by artillery. Lower photo unknown date.

Factory views of Ausf M outfitted with rear reinforced mud flaps.

Pictures of Ausf M modified to utilize what is believed to be a late war fuel substitute, possibly bottled gas. Photos below show components of system.

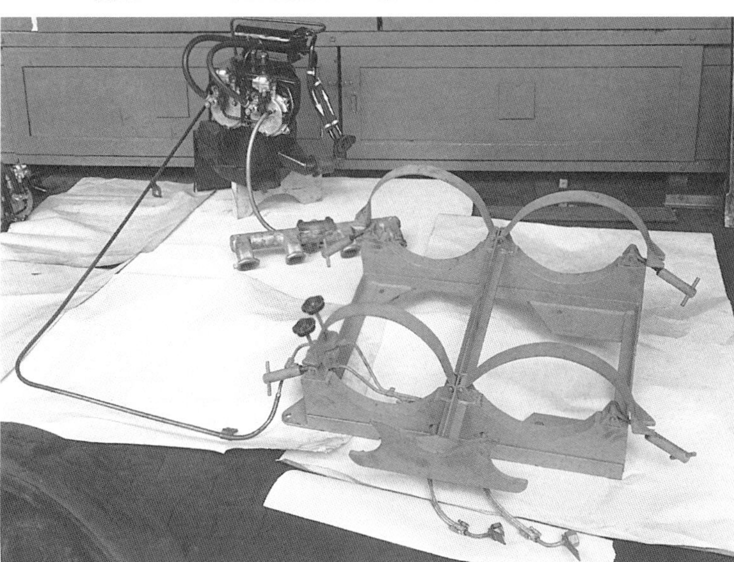

VII 7.5 cm Prototype

This prototype incorporates a 7.5 cm StuK 40(L/43). This gun was produced for the Sturmgeschutz III. The prototype illustrates the design of armored shield which appears to have been used for the 7.5 cm PaK 40/3 Ausf H. The prototype was never introduced into production.

Right side view. Note new style gun shield and fighting compartment housing with wedge shaped mantlet. Driver's vision block not mounted due to angle of fighting compartment armor. Co-ax bow machine gun mount plate missing.

Front left three quarter view.

Right: Front view. Travel lock appears to be a temporary fitting. Note aperture for bow machine gun mount.

Below: Overhead rear view of fighting compartment. White gun shield denotes gun likely removed from late Sturmgeschultz.

Front top view. Aperture on fighting compartment roof for panoramic periscope.

Rear view of same vehicle.

VIII Panzerjager 38(t) fur 7.62cm PaK 36(r) (Sd.Kfz 139)

Vehicle Type: Panzerjager 38(t) mit 7.62cm PaK 36(r) (Sd.Kfz 139)
Manufacturer: BMM
Production Number: 363
Armament: 7.62cm PaK 36(r) 7.92mm MG37
Crew: 4
Weight: 10.67 tons
Length: 5.85 meters
Width: 2.16 meters
Height: 2.5 meters
Armor: 10 - 50mm
Engine Type: Praga AC
Speed: 47km/hr.
Range: 185km

The 7.62cm PaK 36(r) was the first modification to the 38(t) chassis. The vehicle was born from the conversion of the 38(t) chassis and the mounting of captured 7.62 cm Russian anti-tank guns. The fighting compartment, located in the middle of the chassis, surrounded the gun with a superstructure which was open to the rear of the vehicle. The gun shield at 10 mm provided minimum protection. The gun was rechambered for German ammunition, that of the German PaK 40. Thirty rounds of ammunition were carried on the vehicle. The gun shields were reinforced at the rear of the compartment by a 1" diameter crossbar. The seats in place for the gunner, left side, and loader, right side, were of an all metal construction, non-adjustable, and were attached directly to the hull. A wire basket was put in place over the exhaust to retain empty cases after ejection. The gun employed two traveling locks. One of these was located in front of the gun shield with the second being located at the rear of the gun and was comprised of a "U" shaped bracket.

The production run of this vehicle was from April to October of 1942 and 393 vehicles in total were manufactured. This number was split between the Ausf G chassis on which 194 units were built and the Ausf H chassis on which 150 were constructed. An additional 39 vehicles were converted from PzKpfw 38(t) vehicles.

Right side view. Gun at 0 degrees. Travel lock in down position. Travel seats in upright position. Machine gun mounted.

Front left three quarter view. Travel lock in up position. *(Photo Courtesy Imperial War Museum)*

Detail of travel lock. Hinge style cover for recuperator assembly used. *(Photo Courtesy Imperial War Museum)*

Front view. Transmission maintenance hatch clearly visible. Driver and bow gunner vision ports open. *(Photo Courtesy Imperial War Museum)*

Rear view. Note gun shield attachments. Loader's and gunner's seats clearly shown. Also observe rear catch and stowage bracket. Traverse and elevation controls visible. *(Photo Courtesy Imperial War Museum)*

Left rear three quarter view. View of fender reinforcing brackets. *(Photo Courtesy Imperial War Museum)*

Closeup of fighting compartment. Traverse and elevation controls illustrated. Hinged post style seats in place. Breech assembly and guard also visible as well as telescopic sight mount. *(Photo Courtesy Imperial War Museum)*

Closeup of breech. Note detail of assembly saddle and saddle support bracket. *(Photo Courtesy Imperial War Museum)*

Above: Overhead view of fighting compartment. Unique photo illustrating what is believed to be a heating system. Duct work located on inside of fighting compartment. Good detail of support bracket for gun saddle and mounting bracket sockets for seat back on outside of rear stowage basket. Loader's seat hinged and placed over right side of fighting compartment.

Right: Interior driver's compartment. *(Photo Courtesy Imperial War Museum)*

IX 15 cm Schweres Infanteriegeschutz 33(Sf) Ausf H (Sd.Kfz 138/1)

Vehicle Type: 15 cm Schweres Infanteriegeschutz 33(Sf) Ausf H (Sd.Kfz 138/1)
Manufacturer: BMM
Production Number: 90
Armament: 15cm sIG 33/1
7.92mm MG34
Crew: 5
Weight: 11.5 tons
Length: 4.61 meters
Width: 2.26 meters
Height: 2.4 meters
Armor: 15 - 50mm
Engine Type: Praga AC
Speed: 47km/hr.
Range: 200km

This vehicle known as the "Grille" was a self-propelled Howitzer design. The engine was mounted in the rear of the chassis and a fighting compartment superstructure was constructed to encompass nearly the entire length of the vehicle. The front plate of the superstructure extends from the same position as the driver's front plate. The gun was a Rheinmetall-Borsig Howitzer. A sliding armor shield was located below the gun aperture. The driver's position in this vehicle remained as in the 38(t). Construction of the vehicle was done on two separate chassis models, the Ausf K and Ausf M. A total of 90 vehicles were produced between February and April of 1943.

Front left three quarter view. Gun at maximum elevation. Travel lock in up position. Note absence of Notek light.

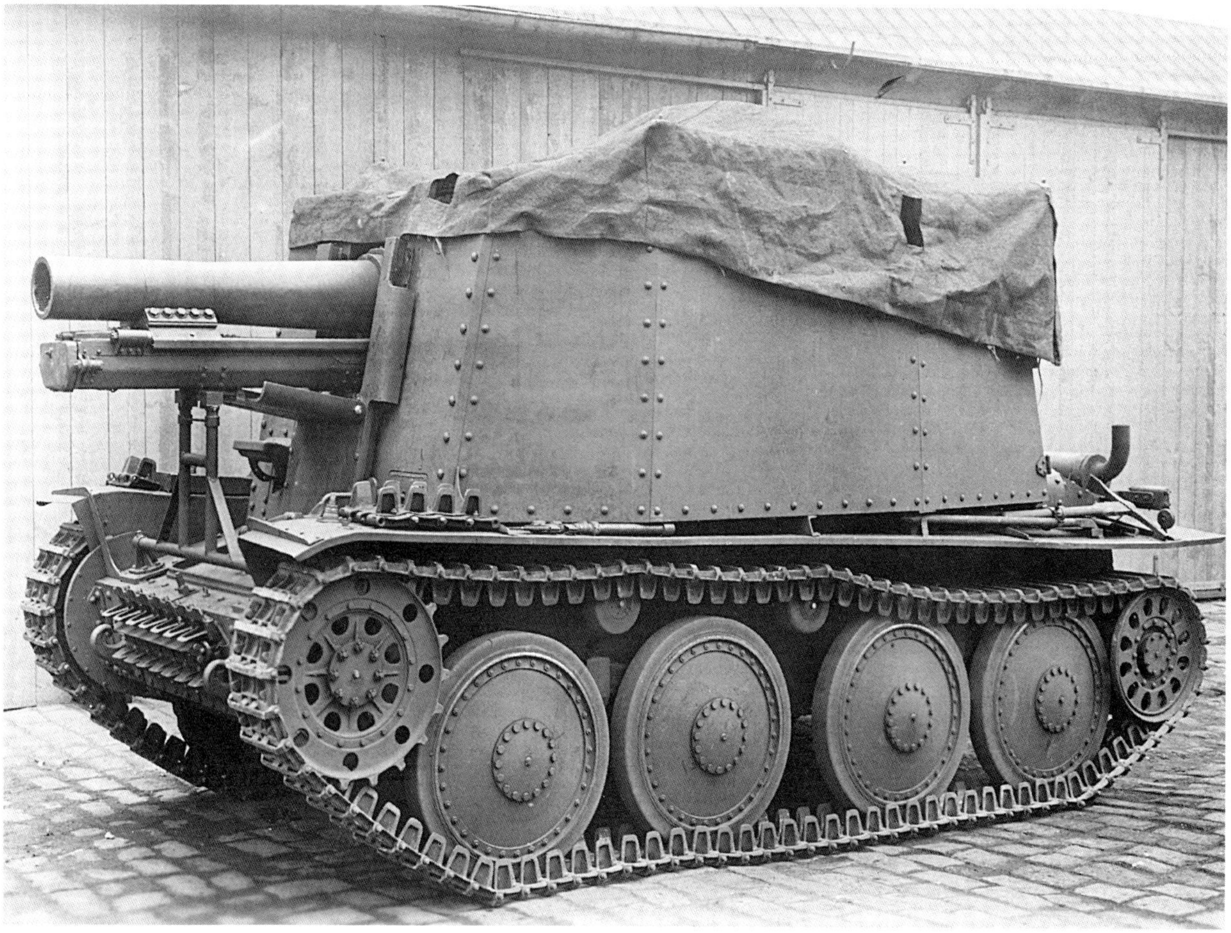

Left side view. Gun at 0 degrees with shield in lowered position. Travel tarp in place.

Right side view. Driver vision devices above jack base visible. Note articulated push rod linkage for travel lock.

Overhead view. Interior fighting compartment details include radio racks with cover as well as ammunition covers. Gunner's seat in lowered position. Elevation hand wheel and mount for periscope visible. Back doors of fighting compartment hinged to outside.

Interior view of driving/radio compartment. Driving controls plainly visible. Note horizontal laterals and control pedals. Clear view of rain guard in place for radios.

Front left three quarter view. Gun in travel lock.

Front view. Gun at max elevation. Aperture for panoramic periscope visible. Push rod linkage for travel lock visible. Driver's vision port open.

Right side view of vehicle being loaded for transport. Left rear door of fighting compartment open showing fire extinguisher. Windows present on canvas cover. Gun barrel cover in place.

Factory shot illustrating fighting compartment structures. Note apertures on fighting compartments. Front vehicle has engine compartment hatches open. Assembly line includes Marder III Ausf H and M's.

X 15cm Schweres Infanteriegeschutz 33/1 auf Selbstfahrlafette 38(t)(Sf) Ausf M (Sd.Kfz 138/1)

Vehicle Type: 15 cm Schweres Infanteriegeschutz 33/1 auf Selbstfahrlafette 38(t)(Sf) Ausf M (Sd.Kfz 138/1)
Manufacturer: BMM
Production Number: 282
Armament: 15cm sIG 33/2
7.92mm MG34
Crew: 4
Weight: 12 tons
Length: 4.95 meters
Width: 2.15 meters
Height: 2.47 meters
Armor: 10 - 50mm
Engine Type: Praga AC
Speed: 42km/hr.
Range: 190km

This vehicle was a continuation of the "Grille" which was constructed on the Ausf M chassis. As with the 7.5 cm PaK 40 Ausf M, the gun was located at the rear of the vehicle, and the engine and the driver were moved towards the front. The superstructure was located behind the driving compartment. The gun was the same as used on the 33(Sf) Ausf H and the front opening when the gun was elevated was protected by a spring loaded hinged armor shield. Two hundred eighty-two of these units were believed to have been produced in two production runs, the first from April to June of 1943, and the second, from October, 1943 to September, 1944. As with the 33(Sf) Ausf H, the vehicle carried a relatively limited amount of ammunition. Eighteen rounds with the Ausf M and sixteen rounds with the Ausf H. The gun used two part ammunition and the fighting compartment provided separate storage for the shells and powder charges. The Ausf M had a hinged rear door which folded down over the muffler and incorporated a shell holder for three rounds.

Front right three quarter view. Gun at full elevation. Note position of spring loaded splash shield. Open padded driver's hatches. Vision hatch closed.

Rear overhead view. Clear illustration of ammo tubes and stowage boxes. Three ready racks are on the lowered rear door. Radios located on rear left hand corner of fighting compartment with antenna located above. Panoramic periscope in place.

Front left three quarter view. Gun in travel lock. Note remnants of wires securing camouflage netting. *(Photo Courtesy Imperial War Museum)*

Right: Front view. Vehicle of 1st SS Leibstandarte Adolf Hitler (LAH). *(Photo Courtesy Imperial War Museum)*

Left: Rear right three quarter view. Note muffler missing. *(Photo Courtesy Imperial War Museum)*

XI Munitions Panzer 38(t)(Sf) Ausf M (Sd.Kfz 138)

Vehicle Type: Munitions Panzer 38(t)(Sf) Ausf M (Sd.Kfz 138)
Manufacturer: BMM
Production Number: 102
Armament: N/A
Crew: 2
Weight: 12 tons
Length: 4.95 meters
Width: 2.15 meters
Height: 2.47 meters
Armor: 10 - 50mm
Engine Type: Praga AC
Speed: 47km/hr.
Range: 200km

Built as a means to supplement the limited ammunition storage of the 33(Sf) Ausf H and 33/1 Ausf M this ammunition carrier was also constructed on the "Grille" chassis. A total of 102 units were manufactured from January to May of 1944. Ammunition racks were installed in the fighting compartment and a total of 40 rounds were carried. An armor plate was put in place over the gun aperture. A ratio of two munition carriers for each six armed Howitzer vehicles was to be used.

Front left three quarter view of early version model. Notek light in place. Driver's hatch padding clearly visible.

Front three quarter view. Front armor removed showing view of compartment. Vehicle still has travel lock for removed gun.

Foward view of vehicle. Note storage frame in compartment.

Rear right three quarter view. Rear armor folded down.

Rear view showing interior storage box and ready racks.

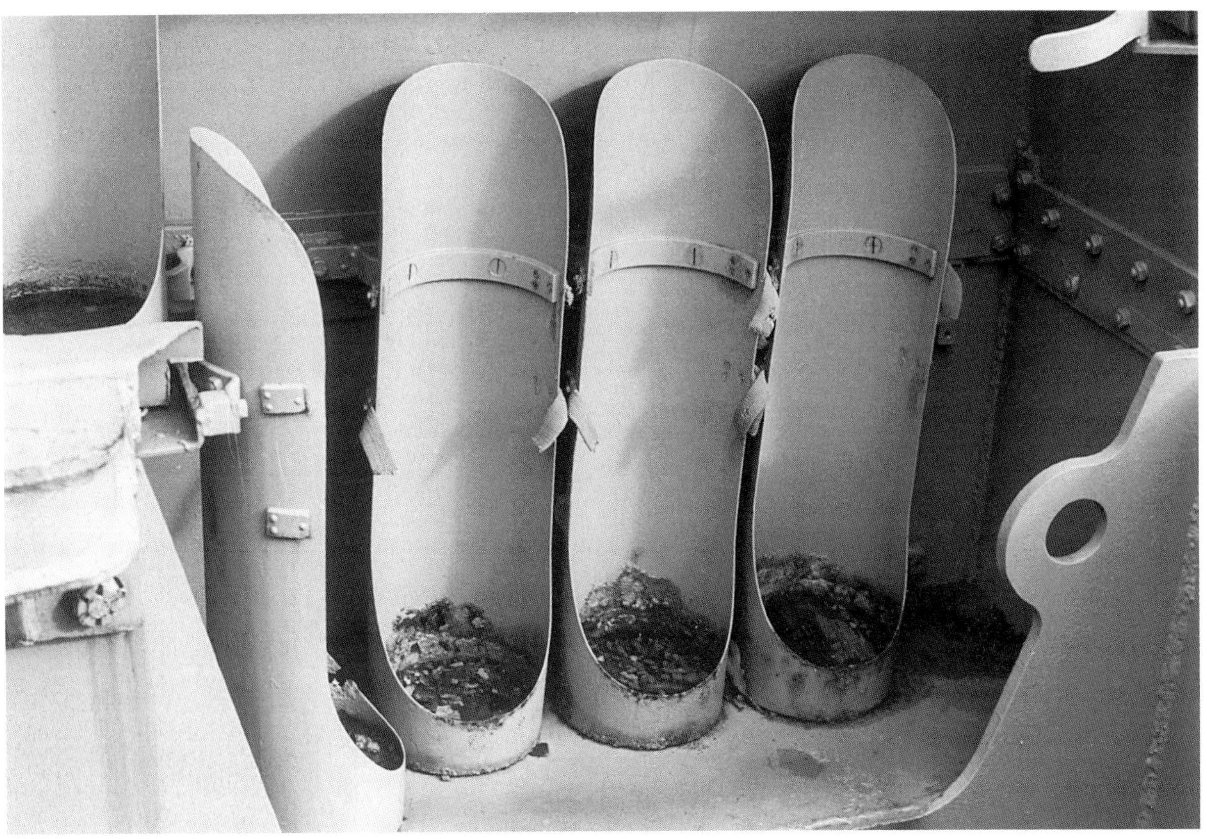

Close up of lower right ammo storage racks.

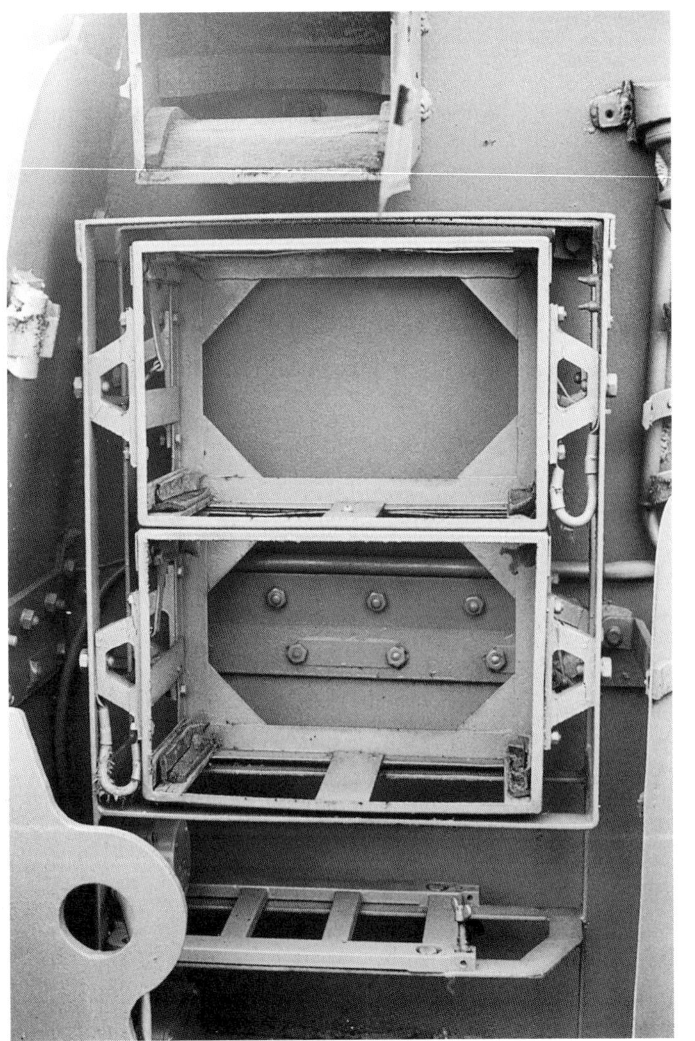

Close up of radio mounts for FU5 and FU7 radios on left side of compartment.

Above: Close up of lower right ammo storage racks. Below: View of ammunition compartment and of radio mounts for FU5 and FU7 radios on left side of compartment.

XII Ausf M Personnel Carrier Prototype

This prototype troop carrier was built on an Ausf M chassis and fitted a one piece armor plate over the former gun port.

Front right three quarter view. Good view of spare track mount.

Right rear three quarter view. Note change in extension of rear superstructure comprising vertical rear doors and step over muffler.

Rear view. Note wooden benches and foot rail. The foot rail blocks PTO hookup as evidenced by plate position. Clear view of vents for what is likely compartment heater. Reinforcement bar mounted across doorway.

Bibliography

Aberdeen Proving Grounds, "Tank Data", WE, Inc. (date unknown).

Chamberlain, Peter & Doyle, Hilary, "Encyclopedia of German Tanks of World War Two" (Arco Publishing, New York 1978).

Doyle, Hilary L. & Kliment, Charles K., "Chechoslovak Armoured Fighting Vehicles 1918-1945" (Argus Books Ltd., Watford 1979).

Kliment, Charles K. & Vladimir, Francev, "Chechoslovak Armoured Fighting Vehicles 1918-1948" (Schiffer, 1997).

McLean, Donald B., "Illustrated Arsenal of the Third Reich" (Normount Technical Publications, 1973).

Scheibert, Horst, "Marder III" (Schiffer, 1998).

Speilberger, Walter J., "Die Panzer-Kampfwagen 35(t) and 38(t)" (Motorbuch Verlag, 1980).

Von Senger und Etterlin, F. M., "German Tanks of World War II" (Galahad Books, 1969).